I0427397

BEEKEEPING BLISS

Mastering The Art Of Apiculture
For Honey And Pollination

Dive Into The World Of
Beekeeping And Harvest The
Sweet Rewards Of Your Own
Honeybee Colony

Dr. Fabian Felicity

Table of Contents

CHAPTER ONE

Introduction

Bees live in a fascinating and complex ecology that is essential for plant pollination and honey production. Beekeeping, or the art and science of caring for bees, has grown in popularity in recent years as more people appreciate the value of these hardworking insects.

Whether you are a seasoned beekeeper or a beginner, learning the fundamentals of beekeeping equipment, hive location, and choosing the appropriate honeybee species is critical for a successful entry into this rewarding activity.

The attractive World of Bees

Bees, frequently regarded as nature's small miracles, are social insects renowned for their exceptional work ethic and intricate hive constructions.

The honeybee, a species of the Apis genus, is most known for its capacity to create honey, a delicious substance that humans have adored for generations. In addition to honey production, bees play an important role in pollinating blooming plants, which helps to reproduce many crops and improves ecosystem health.

Honeybees live in a highly structured colony structure. A normal colony has a single queen, drones (male bees), and worker bees. The queen is the only fertile female capable of depositing eggs, while drones exist only for mating reasons. The bulk of the colony consists of worker bees, which are sterile females who perform numerous activities such as foraging, nursing, and hive maintenance.

Starting with Beekeeping Equipment

Beginning the path of beekeeping necessitates the purchase of vital equipment to guarantee the well-being of both the beekeeper and the bees. The most basic piece of equipment is the beehive, a building that houses the bee colony. Beehives exist in a variety of styles, with the

Langstroth hive being one of the most popular.

The Langstroth hive is made up of stacked boxes, or supers, with detachable frames that contain the honeycomb. This design enables for simple examination and manipulation of the hive while minimizing interruption to the bees. The frames in the supers offer a structured framework for the bees to develop their comb and store honey, making honey extraction a simple operation.

Protective gear is another important component of beekeeping equipment. To avoid bee stings, beekeepers wear specialized outfits that comprise a veil, gloves, and a full-body suit. The veil covers the face and head, while gloves keep hands safe during hive inspections. The full-body suit offers complete protection against possible bee stings, enabling beekeepers to operate safely with their colonies.

Furthermore, a smoker is an important piece of equipment for beekeepers. The smoker emits cold smoke, which calms bees. When used properly, the smoker reduces bee hostility and facilitates hive inspections. Beekeepers may establish an atmosphere in the hive that encourages bees to gorge on honey, making them less inclined to sting.

Choosing the Right Location for Your Beehive

Choosing the best site for your beehive is an important choice that may have a big influence on the success of your beekeeping business. Bees flourish in settings with plenty of nectar and pollen sources, so find a site with a variety of flowering plants. The hive should be located in a location with easy access to water, since bees use water for a variety of tasks, including cooling the hive and diluting honey.

When selecting a place for your beehive, consider the amount of solar exposure. Bees are cold-blooded insects that depend on external warmth to keep their colony at a constant temperature. Placing the hive in a place with morning sunshine is important since it allows the bees to begin their activity early in the day. However, providing some shade during the warmest portion of the day might be beneficial in preventing the hive from overheating.

When choosing a location for your beehive, keep neighbors and municipal rules in mind. While bees are normally non-aggressive while foraging for nectar and pollen, it is critical to locate the hive in a position that avoids possible confrontations with neighbors.

CHAPTER TWO

Choosing The Perfect Honeybee Species

Choosing the correct honeybee species for your beekeeping endeavor requires careful consideration of both your aims and the local environment. Apis mellifera, the most widely maintained honeybee species, is noted for its tolerance to many temperatures and environments. The Apis mellifera species is divided into subspecies, each with its own set of features.

Italian bees, a subspecies of Apis mellifera, are known for their gentle nature, making them an ideal option for new beekeepers. They are prolific honey producers with high resilience to typical hive pests and illnesses. Carniolan bees, on the other hand, are well-known for their winter hardiness and superior foraging ability. These bees thrive in cooler areas.

Buckfast bees, a hybrid developed by Brother Adam at Buckfast Abbey, England, combine qualities from multiple honeybee subspecies. They are known for their tenderness, production, and resilience to illness. Buckfast bees have become popular among beekeepers looking for a well-rounded, versatile honeybee.

In contrast, Africanized bees, sometimes known as "killer bees," are a mix of African and European honeybee species. While Africanized bees are outstanding foragers and honey producers, they are notorious for their aggressive nature, making them unsuitable for beginning beekeepers or populous regions.

Finally, the choice of honeybee species is determined by elements like climate, beekeeping objectives, and the beekeeper's skill level. It is best to check with local beekeeping groups or experienced beekeepers in your region to get insights and advice on the best honeybee species for your unique needs.

Beekeeping is a fascinating trip into the complicated world of bees, delivering a gratifying experience for those who understand the necessity of these pollinators. Understanding the fundamentals of beekeeping equipment, hive layout, and choosing the appropriate honeybee species is critical for maintaining a successful and pleasant relationship with your bee colony.

As you learn about the art and science of beekeeping, bear in mind that each hive is a unique microcosm, with its dynamics impacted by environmental conditions, hive management approaches, and the intrinsic features of the honeybee species you choose.

By approaching beekeeping with knowledge, care, and respect for these hardworking insects, you may begin on a rewarding adventure that not only produces tasty honey but also helps to maintain the health and sustainability of our natural ecosystems.

Building and Setting Up Your Beehive

Beekeeping, an age-old profession, improves the environment while also providing a delicious reward in the form of honey. For those new to beekeeping, the initial step is to construct and set up a beehive. A beehive is the center hub of a honeybee colony, providing housing, room for honey production, and the basis for the bees' complex social organization.

Choosing The Right Location

Choosing an optimal site for your beehive is critical to the success of your beekeeping endeavor. Bees flourish in regions with plenty of sunshine, shelter from severe winds, and access to water supplies.

Placing the hive near a garden or blooming plants guarantees an ample supply of nectar, which is necessary for honey production. Additionally, keep the hive slightly raised to reduce waterlogging and improve hive ventilation.

CHAPTER THREE

Building the Beehive

When building a beehive, there are many designs to select from, including Langstroth, Top Bar, and Warre. The Langstroth hive, with its replaceable frames, is frequently used because of its simplicity and efficiency in honey extraction.

Whatever style you pick, utilizing high-quality, untreated wood is critical to preventing dangerous chemicals from entering the hive. The hive should include a bottom board, brood boxes, supers for honey storage, and a protective roof.

Assemble the Hive Components.

The stability and operation of the hive are dependent on the right assembly of its components. The brood boxes, where the queen lays her eggs, are located at the bottom, while the honey supers, where bees store honey, are piled above.

Frames with wax foundation sheets should be put within the boxes to provide a structure for the bees to construct their combs on. The whole hive should be firmly attached, and careful attention to detail is required to avoid any gaps that might jeopardize the hive's integrity.

Providing Enough Ventilation

A well-ventilated hive is critical to the health of the colony. Proper ventilation aids in the regulation of temperature and humidity inside the hive, reducing the accumulation of excess moisture that may cause mold and sickness.

Ensure that the hive has enough ventilation by using ventilation holes and a screened bottom board. Consider tilting the hive slightly to enable rainfall to drain and lessen the chance of waterlogging.

Understanding The Life Cycle Of Honey Bees

To succeed as a beekeeper, you must first understand the honeybee life cycle. Honeybees have a specific life cycle, with each stage playing an important part in the colony's survival and output.

The Queen Bee

The queen bee is important to every honeybee colony. The queen is the colony's single reproductive female and is in charge of depositing eggs. She may lay up to 2,000 eggs in a single day, ensuring that the colony grows and replenishes itself. The queen's pheromones also help to preserve the hive's social cohesiveness, giving the worker bees a feeling of belonging.

Worker Bees

Female worker bees make up the vast bulk of a colony's population. Throughout their lives, worker bees take on several duties, beginning as cleaners, progressing to nurse bees, and eventually becoming foragers. Their responsibilities include feeding the queen and larvae, creating combs, and hunting for nectar and pollen. Worker bees are the hive's backbone, ensuring its general operation and production.

CHAPTER FOUR

Drone Bees

Male bees, known as drones, have a single purpose: to mate with a queen from another colony. Drones do not engage in hive operations like foraging or nursing. Their presence is critical during the mating season, after which they perish. Drones serve an important role in preserving genetic variation among honeybee populations.

The Egg, Larva, And Pupa Stages

The queen deposits an egg in a comb cell, beginning the life cycle. The egg develops into a larva, which the worker bees feed and care for. The larva subsequently undergoes metamorphosis, becoming a pupa. During this stage, the cell is shut and the pupa grows into an adult bee. The time of this cycle varies by kind of bee, with drones taking the longest to mature.

Beekeeping Seasons: Managing Your Colony Throughout The Year

Beekeeping is a year-round activity, and knowing seasonal patterns is critical to properly maintaining a honeybee colony. Each season brings new difficulties and activities for beekeepers to guarantee the health and production of their bees.

Spring

Spring signals the start of increasing hive activity. As temperatures increase, bees become more busy in their quest for nectar and pollen. This is also the time for swarming when a new queen and some of the colony leave the hive to start a new one. Beekeepers must be attentive in the spring, examining hives for indications of swarming and giving more space if needed.

Summer

Summer is the prime season for honey production. With plenty of feed available, bees work hard to fill honey supers. To avoid overpopulation and swarming, beekeepers should check for full supers regularly and gather honey. It is also critical to offer enough water sources for the bees during the hot summer months.

Fall

As temperatures drop, bees divert their attention away from feeding and toward winter preparation. Beekeepers should minimize the number of hive openings to keep pests and predators out. Furthermore, autumn is the period to evaluate the hive's food reserves. If honey supplies are inadequate, the colony may need additional feeding to survive the winter.

Winter

Winter is the biggest threat to honeybee hives. To keep warm, bees construct a winter cluster, with the queen in its core. Beekeepers must maintain the hive well-insulated and sheltered from extreme weather conditions.

During the winter, it is critical to monitor food supplies and provide additional feeding as needed. Regular hive inspections are restricted to prevent causing excessive disruption to the cluster.

Keeping A Healthy Beehive: Disease Prevention And Treatment

Maintaining the health of a beehive entails aggressive illness prevention and effective treatment when required. Honeybee colonies are vulnerable to a variety of illnesses and pests, and beekeepers play an important role in mitigating these risks.

Regular Hive Inspections

Regular hive inspections are the primary line of protection against illnesses and pests. During inspections, beekeepers should search for indicators of illness, such as aberrant brood patterns, discoloration, or strange worker bee behavior. Early detection allows for rapid action and reduces disease transmission within the colony.

CHAPTER FIVE

Varroa Mite Management

Varroa mites are a major hazard to honeybee colonies since they prey on both adult bees and growing broods. Beekeepers should use effective varroa mite control measures, including mite-resistant bee varieties, screened bottom boards, and organic treatments like formic acid or thymol.

Disease Prevention Via Hygiene

Maintaining a clean and sanitary hive habitat is critical to illness prevention. To avoid pathogen formation, scrape out the beeswax and propolis buildup regularly and replace the old comb. Proper cleanliness minimizes the likelihood of bacterial and fungal illnesses in the hive.

Supplemental Food And Nutrition

Ensuring that bees have proper nourishment is critical for disease resistance. During times of shortage, beekeepers may need to supply extra nourishment, such as sugar syrup or protein patties, to strengthen the colony. Bees that are well-nourished are more resistant to illnesses and pests.

Integrated Pest Management

An integrated pest management (IPM) method entails combining a variety of tactics to control pests and illnesses rather than depending exclusively on chemical treatments. This involves maintaining healthy colonies, keeping a clean environment, and utilizing organic treatments sparingly when required.

IPM helps to avoid the establishment of pest-resistant strains and decreases the environmental effect of chemical treatments.

Finally, beekeeping is a pleasant and complex hobby that demands meticulous attention to detail and a thorough study of the honeybee lifecycle, seasonal dynamics, and disease control. Building and establishing a beehive is the first phase, followed by ongoing efforts to sustain the colony throughout the year. With appropriate care and attention, beekeepers may enjoy the delicious rewards of their work while also contributing to bees' vital role in pollination and ecosystem health.

Harvesting Honey: Techniques And Tips

Honey harvesting is the satisfying conclusion to the complex and intriguing process of beekeeping. The meticulous harvest of honey from beehives requires a combination of knowledge, expertise, and respect for these hardworking pollinators. Understanding the strategies and suggestions for extracting honey is critical to maintaining a successful and sustained beekeeping operation.

Beehive Inspection Is The Foundation For Successful Harvesting

Before beginning the honey extraction procedure, do a thorough check of the beehive. Regular hive inspections not only help beekeepers evaluate the general health of the colony, but they also offer information on the honey production status. During these examinations, beekeepers may examine the colony's strength, the quantity of capped honey, and the general quality of the comb.

CHAPTER SIX

Timing Is Key: Knowing When To Harvest.

Choosing the appropriate time to gather honey is a tricky balance. Beekeepers must be sensitive to the natural cycles of the colony and the surrounding plants.

Honey should be collected when the bulk of the cells are capped, suggesting that it has attained the proper moisture content.

Harvesting too early may result in honey with too much moisture, whilst waiting too long might cause extraction issues and possibly honey fermentation.

Gentle Harvesting Techniques: Minimizing Stress For Bees

Once the best harvesting period has been determined, using gentle procedures is critical for reducing stress on the bees. Smoke, a classic beekeeping tool, may help calm the bees and make the extraction procedure go more smoothly.

Furthermore, utilizing bee escapes or fume boards might persuade bees to leave the honey supers, allowing beekeepers to gather honey without upsetting the whole colony.

Tools Of The Trade: Extractors And Uncapping Devices

The extraction procedure includes removing honey from the comb, which requires the use of specialized instruments. Honey extractors, both manual and electric, are routinely used to spin honey out of frames.

Beekeepers use uncapping instruments like hot knives or uncapping forks to open the cells and liberate the honey. Proper use of these instruments enables an efficient and productive honey-gathering process.

Processing and Storing Your Honey

Harvesting honey is just the beginning; additional processing and storage stages are equally important in preserving the finished product's freshness and taste.

Uncapping And Extracting: Precise Methods For Quality Honey

After harvesting, the frames must be opened to release the honey. Beekeepers may select between hot and cold uncapping procedures, each with benefits. Hot uncapping uses hot blades, which is faster but may alter the taste of the honey. Cold uncapping, on the other hand, protects the honey's inherent properties while requiring more time and work. Once uncapped, the frames are put in an extractor to collect the honey while keeping the empty comb intact for the bees to reuse.

Filtration: Refining Liquid Gold

Honey is filtered to produce a clear and visually pleasing end product. This phase eliminates any leftover contaminants or particles, leaving a smooth, transparent liquid. While some beekeepers prefer minimum filtering to maintain the honey's natural properties, others use finer filters to obtain a more polished look.

Proper Storage Preserves Flavor And Quality

Honey's taste and quality must be preserved via proper storage. Honey is hygroscopic, which means that it absorbs moisture readily. Honey should be kept in a cold, dry area to avoid crystallization and keep it liquid.

Using airtight containers protects it from external factors and increases its shelf life. Glass jars are a popular option for preserving honey since they are non-reactive and do not impart any unpleasant flavors.

CHAPTER SEVEN

Exploring Bees' Role in Pollination

Aside from the delicious reward of honey, bees serve an important role in the ecology as pollinators. Understanding this ecological function highlights the need to conserve and sustain bee populations.

The Pollination Process: Nature's Symbiotic Relationship

Pollination is a natural process that transfers pollen from a flower's male to female sections, allowing for fertilization and seed formation.

41

Bees, by their foraging habit, unwittingly transport pollen from blossom to flower as they gather nectar. Bees are among the most successful pollinators in nature because of their inadvertent but critical involvement in plant reproduction.

Biodiversity And Food Production: Bees' Effect On Agriculture

The agricultural industry depends significantly on bees for pollination, which improves crop productivity and quality. Bees pollinate a wide variety of crops, vegetables, and nuts, including apples, almonds, and berries. The loss of bee populations presents a huge danger to the world's food supply and biodiversity, highlighting the need for conservation initiatives and sustainable beekeeping methods.

Conservation and Sustainable Beekeeping: Preserving Bee Habitats

As the importance of bees in pollination becomes more widely recognized, there is a greater focus on conservation and sustainable beekeeping techniques. Preserving natural habitats, reducing pesticide usage, and supporting measures to safeguard bee populations are critical steps toward preserving ecological equilibrium. Educating communities on the significance of bees and their influence on food security develops a shared responsibility to preserve the survival of these vital pollinators.

Creating A Bee-Friendly Garden

Incorporating bee-friendly components into gardens not only benefits the health of these important pollinators but also improves the general attractiveness and productivity of outdoor areas.

Choosing Bee-Friendly Plants: A Diverse Menu For Bees

Creating a bee-friendly garden starts with choosing plants that attract and feed bees. Choosing a varied selection of blooming plants with various bloom periods ensures that bees have a consistent supply of nectar and pollen throughout the seasons. Native plants are especially well-suited to attracting local bee species since they have developed together over time.

Avoiding Harmful Chemicals: Pesticide-Free Gardening

To maintain a bee-friendly garden, avoid using dangerous chemicals and pesticides. Bees are particularly sensitive to manmade pesticides, and their usage may hurt bee populations. Practicing organic gardening and using natural pest control methods helps to create a safe and welcoming environment for bees to flourish in.

Providing Shelter And Water: A Complete Habitat

In addition to feeding, bees need appropriate shelter and water supplies. Including features like bee homes, which offer nesting locations for solitary bees, and shallow water containers with landing platforms ensures that bees have all they need for a healthy environment. A well-balanced garden that recognizes the holistic requirements of bees helps to improve the general health of local bee populations.

Finally, beekeeping is much more than just extracting honey. It entails a thorough grasp of bee behavior, a dedication to sustainable methods, and an awareness of the critical role bees play in pollination. As beekeepers master the procedures of honey extraction, processing, and storage, they contribute not just to the sweetness of honey but also to the larger ecological balance that keeps our world alive. Simultaneously, developing bee-friendly places in our gardens promotes a peaceful connection between people and these hardworking pollinators, assuring their continuing success.

CHAPTER EIGHT
Tackling With Common Beekeeping Challenges

Beekeeping, is a joyful and ecologically vital pastime, with its own set of obstacles. Successful beekeeping requires a thorough grasp of hive dynamics and the ability to handle typical problems.

Whether you're a first-time beekeeper or an experienced apiarist, dealing with problems as soon as possible preserves the health of your colony and apiary production.

Identifying Health Concerns In The Hive

One of the most difficult aspects of beekeeping is identifying and resolving health concerns inside the hive. Varroa mites, tiny parasitic insects, are a major danger to honeybee hives. Regular hive checks are critical for detecting the presence of these mites early on.

Monitoring for symptoms of brood diseases, such as American Foulbrood or European Foulbrood, is also important. Timely action, including suitable treatments and, if required, culling sick colonies, aids in the prevention of disease transmission.

Varroa Mites Provide A Persistent Threat

Varroa mites adhere to adult bees and larvae, feasting on their body fluids and spreading viruses. Integrated Pest Management (IPM) tactics, such as employing screened bottom boards and organic acids, may aid in mite management. Beekeepers should be watchful and apply these procedures to ensure a healthy bee population.

Navigating Seasonal Challenges

Beekeeping is a dynamic activity that is impacted by the seasons. Challenges change throughout the year, necessitating unique tactics for each season. Winter provides hazards like hunger and cold stress, while spring raises the possibility of swarming as the colony grows.

Adequate planning, such as giving supplemental nourishment in the winter and monitoring hive space in the spring, may help to reduce these issues.

Winter Survival Strategies.

Monitoring food stocks is critical to ensuring your colony's survival over the winter. Bees need enough honey stores to make it through the winter months when foraging becomes difficult. If necessary, additional feeding with sugar syrup or fondant might offer the required energy. In addition, winter hive management requires insulating the hive and providing sufficient ventilation.

Swarming Prevention in the Spring

Spring is a time of development and expansion for bee colonies, which often leads to swarming. Beekeepers may avoid swarming by allowing enough room for the expanding population. Regular hive inspections and the installation of supers (extra hive boxes) may assist manage space and reduce the possibility of swarming. Understanding the swarm impulse and implementing preventative actions are critical to sustaining a stable colony.

CHAPTER NINE
The Business Of Beekeeping: Making Your Hobby A Profitable Venture

For many beekeepers, what starts as a pastime may turn into a lucrative business. Turning your love for beekeeping into a profitable company involves strategic planning, market insight, and a dedication to environmentally friendly techniques.

Individuals who embrace the commercial side of beekeeping may contribute not only to the health of their colonies but also to their financial success.

Market Analysis And Product Diversification

Before starting a beekeeping company, you need to undertake a complete market study. Understanding local demand, pricing tactics, and prospective competition enables beekeepers to properly market their goods.

Furthermore, expanding product options, such as selling honey, beeswax candles, or pollen, will help you reach a larger consumer base and generate more cash.

Branding And Marketing Strategy

A successful beekeeping company requires the development of a brand identity as well as efficient marketing methods. Creating a unique selling proposition (USP), such as supplying organic or locally sourced honey, may help a firm stand out in a competitive market.

Using internet platforms, social media, and local farmers' markets may help you reach a larger audience and develop a loyal consumer base.

Sustainable Beekeeping Practices: Greener Apiaries

As environmental concerns develop, beekeepers are increasingly prioritizing the health of bees and the habitats in which they live. Sustainable beekeeping entails reducing environmental impact, encouraging biodiversity, and maintaining long-term health for both bees and their surroundings.

Organic Beekeeping Methods

Choosing organic beekeeping practices means avoiding synthetic chemicals and pesticides. Instead, beekeepers employ natural alternatives and integrated pest control strategies to solve hive health concerns. This strategy not only protects bees from dangerous toxins but also helps to produce high-quality organic honey, which is sought after by ecologically aware customers.

Enhancing Bee Habitats

Creating a bee-friendly environment extends beyond the apiary. Beekeepers may help to increase biodiversity by planting pollinator-friendly plants and flowers in their surroundings. Creating a broad foraging environment means that bees have access to a range of nectar and pollen sources, which improves their general health.

Conclusion

Beekeeping, in addition to being a rewarding pastime, has several problems and possibilities. Getting through typical beekeeping issues takes a mix of knowledge, attentiveness, and proactive management.

As beekeepers attempt to transform their pastime into a lucrative business, the incorporation of sustainable methods becomes critical.

Beekeepers play an important role in protecting these key pollinators for future generations by adopting ecologically friendly practices and contributing to their well-being.

Whether solving hive difficulties or investigating the commercial elements of beekeeping, adherence to sustainable methods fosters a peaceful interaction between beekeepers and the fascinating world of honeybees.

www.ingramcontent.com/pod-product-compliance
Lightning Source LLC
Chambersburg PA
CBHW072257310526
45795CB00012B/1711